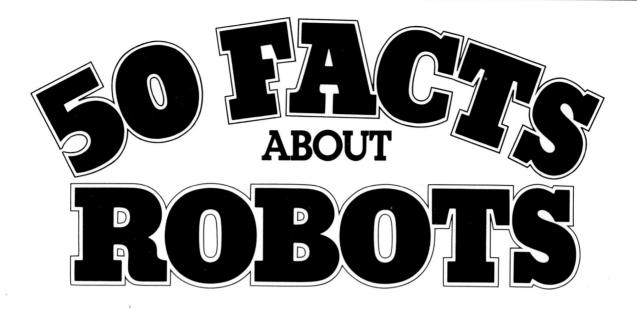

50 FACTS ABOUT ROBOTS

by
Mark Lambert

WARWICK PRESS

CONTENTS

		page			page
1	How did robots begin?	3	28	How was Mars explored?	17
2	What is a robot?	3	29	What does Sim One do?	18
3	Is an automaton a robot?	4	30	Is Bionic Man possible?	18
4	What is feedback?	4	31	How do artificial arms work?	18
5	What makes a real robot?	5	32	What is an exoskeleton?	19
6	Where do you find robots today?	5	33	What is the Walking Lorry?	19
7	What are performing robots?	6	34	What are second generation robots?	20
8	What is an android?	6	35	How can a robot see?	20
9	What type of robot is a cyborg?	6	36	How can robots feel?	21
10	Are robots dangerous?	7	37	Can you talk to a robot?	21
11	What are the Laws of Robotics?	7	38	Can robots walk about?	21
12	What is a robot's brain?	8	39	Can a robot brain think?	22
13	How does a robot's brain work?	8	40	What can WABOT do?	22
14	What is a robot's language?	9	41	How clever was Shakey?	23
15	How does a robot know what to do?	9	42	Are robot mice intelligent?	23
16	How does a robot arm work?	10	43	Could robots run factories?	24
17	What is a first generation robot?	10	44	Could robots work on farms?	25
18	What is a pick and place robot?	11	45	How will robot cars work?	26
19	How do robots help make cars?	11	46	Will we have robots at home?	27
20	Does a robot know what it's doing?	12	47	Could we have robot pets?	27
21	How do you teach a robot?	12	48	Will we use robot weapons?	28
22	Can a robot learn to paint?	13	49	Could robots repair spacecraft?	28
23	What is remote control?	14	50	Could robots build space cities?	29
24	Why is remote control useful?	15		**Robot Maze Games**	30
25	Can we feel by remote control?	15		**The Multiple Robot Game**	31
26	What does Wheelbarrow do?	16		**Index**	32
27	Did robots explore the Moon?	17			

Published in 1983 by
Warwick Press,
730 Fifth Avenue,
New York, New York 10019.
First published in 1983 by
Pan Books Ltd., London

Designed and produced by
Piper Books Ltd., London.
Copyright © by Piper Books. 1983

Printed and bound by
Graficas Reunidas, S.A.,
Madrid, Spain.
All rights reserved.

Library of Congress
Catalog Card No. 83-60604
ISBN 531-09218-6

1 How did robots begin?

The word 'robot' was first used by the Czechoslovakian writer, Karel Čapek (pronounced Chapek). In 1922, he wrote a play called R.U.R. (Rossum's Universal Robots). In this play, artificial humans are made in a factory on a remote island. They are made of metal and have no human feelings. Rossum's Robots are sold all over the world to do work for humans – the Czech word *robota* means 'slave-like work'. Eventually, the robots rebel against their masters and take over the world.

C-3PO (*Star Wars*)

PRAB (Mechanical manipulator)

The Scribe (automaton)

2 What is a robot?

The word robot is used to describe a number of very different things. To many people a robot is any machine that physically looks like a human being, or even an animal. These are the kind of robots that appear in stories, like Čapek's play, and in science fiction films. Sometimes even automata are thought of as being robots, even though they are really only mechanical toys.

In the real world today, a robot is a very different thing. Modern industrial robots do not look like human beings. But they often perform very human-like actions; their 'arms' may move in a similar way to human arms, for example.

Remote controlled machines can also be thought of as robots, as they often have many robot-like abilities. Of course, unlike industrial robots, these machines have to be operated directly by a human being. However, when a human operator is very closely linked to a remote controlled machine, as in the case of the Walking Lorry (see page 19), the human/machine partnership becomes almost like a cyborg (see page 6).

3 Is an automaton a robot?

The Lute Player (automaton)

For over two thousand years people have tried to build mechanical objects that looked and moved like humans or animals. The most successful of these machines were made in the 1700s and 1800s and are known as 'automata'. The Scribe, shown on the previous page, appears to write in a very human-like way. While the Lute Player, in this picture, walks and moves her head from side to side as she plucks at the lute strings.

However, like any clockwork machine, an automaton can only perform a single set of actions. It has no brain and cannot change any of its movements. The modern idea of a robot has advanced beyond this stage.

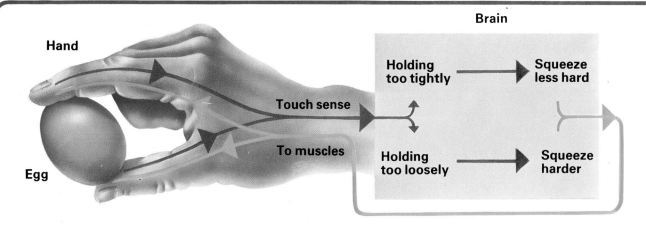

4 What is feedback?

An automaton does not have a brain. A robot does – its brain is a computer. But a computer by itself is still not enough to make a robot do things in a human-like way. It also needs to have something called 'feedback'.

Feedback controls everything you do – every action you make. Information about what is happening around you is collected by your eyes, ears and other senses. This information is fed back to your brain and either directly controls, or, at least, helps to control what you do next.

Think about a simple action like holding an egg. Information about how tightly you are holding the egg is fed back to your brain by the sense of touch in your fingers. If you are squeezing too hard and are likely to crush the egg, your brain sends out a message to your finger muscles to make them relax. If your fingers become too relaxed and you are in danger of dropping the egg, your brain tells your finger muscles to squeeze harder. Thus, feedback makes sure that you are holding the egg in just the right way.

What makes a real robot?

A true robot must be as human-like as possible. So, first, it must be able to do something in a similar way to a human. However, it does not necessarily have to look like a human in order for it to carry out its tasks efficiently.

Second, it must have a brain, in the form of a computer. Third, it must have feedback, that is, it must have some senses, such as sight, hearing, or a sense of touch. Only then will it be able to 'notice' and react to what is going on around it.

Finally, a robot must be able to learn. At the moment, a modern robot 'learns' by having a program – a set of instructions on how to do a particular task – put into its computer.

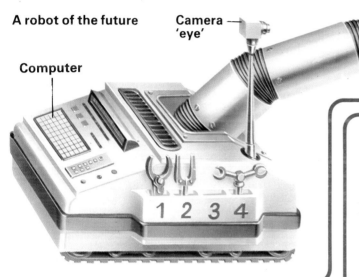

A robot of the future
Camera 'eye'
Computer
Robot arm

Where do you find robots today?

Modern robots are nothing like the spectacular beings of science fiction. However, they are highly advanced machines. Nearly all modern robots are found in factories, where they are used to carry out some of the more boring or heavier work that people do, such as lifting, painting, welding and simple assembly work.

An industrial robot generally has one arm, controlled by a computer brain. Its program can be changed so that it can do a number of different jobs. But modern robots are very stupid; they cannot react to what is happening around them. For example, a robot may go on lifting things even if there is nothing there to lift!

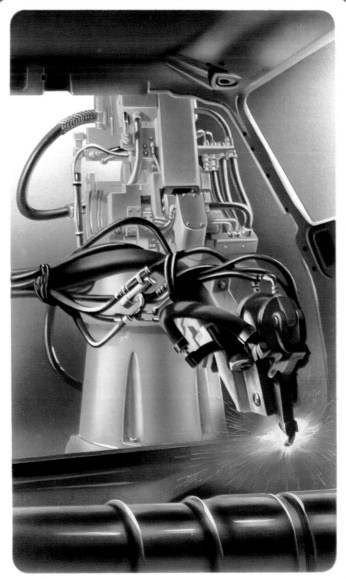

What are performing robots?

Since the early 1900s, some robots have been seen in fairs and public shows. These metal men are called performing robots. They certainly look like science fiction robots, but, in fact, they are all electrically operated and have none of the abilities of real robots. ON/OFF, shown here, guides visitors to the collection of toy robots in the Wonders of the World Museum in California.

What is an android?

An android is a type of robot that, from the outside, looks exactly the same as a human being. Some androids are built from mechanical parts and are then covered with realistic-looking skin and clothes. Like other robots, they can be repaired when necessary, and some can even repair themselves. Other androids are 'grown' from chemicals. Of course, androids exist only in science fiction.

Rem (*Logan's Run*)

What type of robot is a cyborg?

A cyborg is part human and part machine, like the television hero, The Bionic Man, who has bionic limbs and eyes. The name 'bionic' is made up from the two words *bio*logical and electro*nic*. It is used to describe an electronic machine that does the work of any of the parts of a living body.

Other cyborgs include the Cybermen, shown here, from the BBC television series *Dr Who*. These cyborgs were once men. As they became old their human parts were gradually replaced by robot parts, until even their brains were replaced by computers.

10 Are robots dangerous?

Robots are often thought of as very sinister beings. In many films they are shown as frightening creatures who want to wipe out the human race. The Daleks of *Dr Who* are well known examples of such unfriendly robots. But unless robots actually do come from outer space the only robots on Earth will be built by humans. So they need only be as dangerous as we make them.

11 What are the Laws of Robotics?

In a number of films and books there are robots whose aim is to serve and please humans. The robot in the film *Tobor the Great* (1954), for example, was programmed to take special care of children. Robots of this type generally obey the three Laws of Robotics that were created by the writer, Isaac Asimov, in the 1940s:

1. A robot may not injure a human being, or, through inaction, allow a human being to come to harm.
2. A robot must obey the orders given it by human beings, except where such orders would conflict with the First Law.
3. A robot must protect its own existence as long as such protection does not conflict with the First or Second Law.

These laws work quite well in science fiction, but, at the moment, no real robot is intelligent enough to understand them.

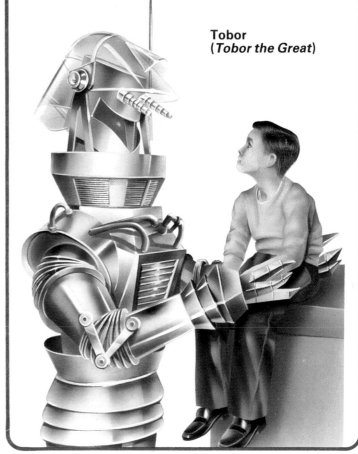

Tobor (*Tobor the Great*)

12 What is a robot's brain?

The brain of a robot is a digital computer. This type of computer works with numbers, or digits. It changes every piece of information it handles into a number code.

A computer has to be fed information to start with. This is often done by an operator, who types it onto the keyboard of a computer terminal. A computer can hold an enormous amount of information and can work things out incredibly quickly.

The first computers were huge. They were made of hundreds of valves and wires. Then transistors replaced valves and computers became smaller. Today, transistors and other electronic parts can be put onto tiny pieces of silicon. These silicon chips are set into plastic cases and plugged into circuit boards. A number of circuit boards may be put together to make a computer.

Computer terminal
Keyboard

Silicon chip

Circuit board

13 How does a robot's brain work?

A robot's brain, or computer, is made up of several main parts. First, there is the input unit, into which information (data) is fed and through which the computer is told what to do. The input unit may be a keyboard, or some other special piece of equipment.

Inside the computer the central processing unit (CPU) receives the input data and sends it to other units. The arithmetic unit does whatever work is necessary, and the memory supplies any additional information that is needed. Sometimes a computer needs information from an extra memory.

When the computer has done its work, the CPU sends the results to the output unit. In a robot the output would be in the form of electrical signals – to make the robot arm move, for example.

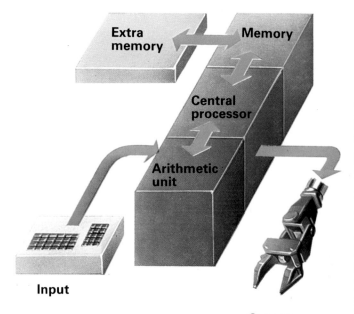
Extra memory · Memory · Central processor · Arithmetic unit · Input · Output

14 What is a robot's language?

Our usual system of counting is called the decimal system. We use ten digits (0 to 9) and when we write out a number, each digit has ten times the value of the one on its right. That is, the number 536 means $(5 \times 100) + (3 \times 10) + (6 \times 1)$.

A digital computer, however, works by using just two digits, 1 and 0. This two-digit system of counting is called the binary system. When a number is written out, each digit has twice the value of the one on its right. That is, the number 1101 means $(1 \times 8) + (1 \times 4) + (0 \times 2) + (1 \times 1)$, which in decimal numbers would be 13.

Although the binary system produces much longer numbers than the decimal system, it is easier for the computer to use. The 1s and 0s can be simply represented by pulses of electrical current, which are controlled by turning switches either on or off. This binary system can be used as a code for letters, colours and shapes as well as numbers. The code can then be recorded on a tape or a disc, or punched onto paper tape by means of 'holes' or 'no holes'.

Binary system of counting

	64	32	16	8	4	2	1
1							1
2						1	0
3						1	1
4					1	0	0
5					1	0	1
6					1	1	0
7					1	1	1
8				1	0	0	0
9				1	0	0	1
10				1	0	1	0

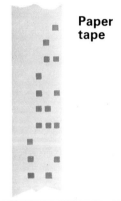

Paper tape

15 How does a robot know what to do?

A robot is told what to do by a set of instructions called a program, which is put into its computer. But because all a computer does is to turn its switches on and off, a program has to break down every general instruction into a series of very simple stages. There are two kinds of stages. An instruction stage tells the computer to perform a straightforward calculation, or to move a robot part. A decision stage is when the computer has to ask itself a question, to which the answer must be yes or no.

The first step in writing a program is to work out a flowchart. The chart drawn here shows just some of the stages involved in fetching a glass of water. To make a robot actually do this would take much more detailed stages than these. Then, a computer programmer must write out each stage in the correct words and figures to be fed into the robot's brain.

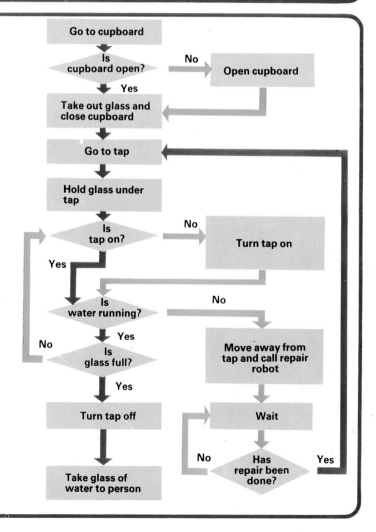

16 How does a robot arm work?

A human's arm and hand are amazingly flexible. Together they have over 40 'degrees of freedom' — that is, different types of movement. In contrast, the most advanced artificial arms (see page 18) have only 10 degrees of freedom. And most robot arms have only six. However, this is generally enough to make a robot arm useful.

As shown in the picture, each degree of freedom in a robot arm is provided by a separate joint. By moving two or more different joints at the same time, the robot arm can move in almost any direction, and can be positioned where it is needed.

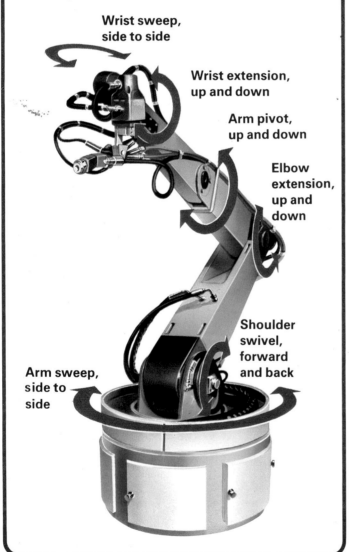

Wrist sweep, side to side
Wrist extension, up and down
Arm pivot, up and down
Elbow extension, up and down
Shoulder swivel, forward and back
Arm sweep, side to side

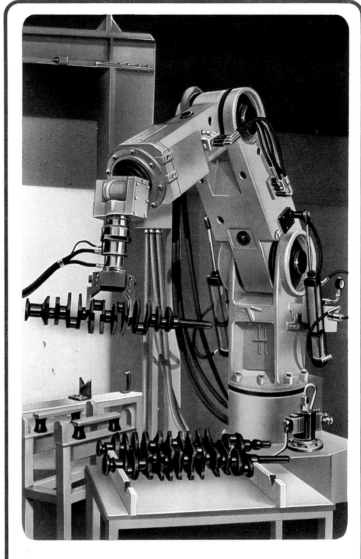

17 What is a first generation robot?

The first robots to be used in industry are now known as 'first generation robots'. These robots are large, heavy machines used for such work as handling hot pieces of metal, loading, welding car bodies and spray painting.

First generation robots have a special tool, such as a gripper or welding gun, fixed to the end of their arms. They are fairly clumsy machines and cannot do very complicated work. But, simply by changing its tool and reprogramming a first generation robot's brain, it can be given any one of a number of different jobs to do.

18 What is a pick and place robot?

One of the simplest – and clumsiest – types of robot is a pick and place robot. This type, as its name suggests, is used to pick objects up from one work area and move them across to another. Using a robot to do this makes a boring job quick and easy.

The movements of the robot's arms are controlled electrically. A control unit makes sure that the arms move in the right direction and in the right order. Mechanical stops on the arms prevent them moving too far.

Programming a pick and place robot is fairly simple. It is done by plugging electrical connections into a plugboard in the robot's control unit. Reprogramming it just means changing the connections.

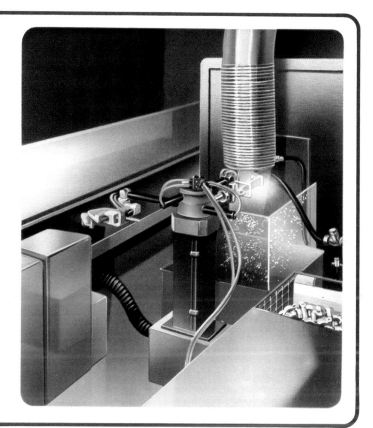

19 How do robots help make cars?

Most cars are mass-produced on production lines. As a car is built, it moves along the line, stopping at different stations for different jobs to be done.

It is easy to introduce robots into such a system. Today, parts of some car production lines are 'manned' entirely by robots. They are mainly used for welding the bodies of the cars together. At the first welding station the body parts are clamped together and robots apply spot welds. Then the clamp, or 'gate', is lifted away and the car body moves on to several more stations where other robots finish welding the body together. By the time a car reaches the end of the welding line over 400 welds may have been made.

The car body is carried along the line on a transporter. Its movement around the factory is controlled by a central computer. Stations that break down or are being used for other work can be passed by.

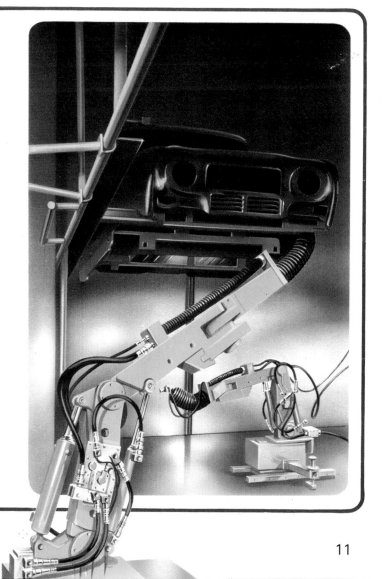

20 Does a robot know what it's doing?

The arm of a pick and place robot swings violently until it crashes into its mechanical stop. The robot has no senses and so cannot know where its arms are, or how fast they are moving.

More advanced robots have feedback and computer memories. A welding robot, for example, has position sensors in each of its joints. These continuously feed back information to the robot brain. As a result, the robot always knows the position of its arm and how fast each part is travelling.

When a joint reaches a position recorded in the robot's memory, the arm stops for a moment before beginning the next movement. The robot follows its program exactly, moving the parts of its arm at a steady speed, according to its instructions.

21 How do you teach a robot?

An industrial robot is taught what to do by linking it up to a teach control unit. Say, for example, the robot is being taught to pick up a metal object and place it carefully in a stack. A human operator presses buttons on the teach unit to move each of the robot's joints slowly into position so that the gripper can pick the object up. When the operator is satisfied that exactly the right position has been reached, he presses a 'record' button.

The operator takes the robot through each movement, pressing the 'record' button at every stage so the positions are fixed in the robot's memory. When the teach unit is removed, the robot will move its arm and gripper to each recorded position in exactly the same way each time.

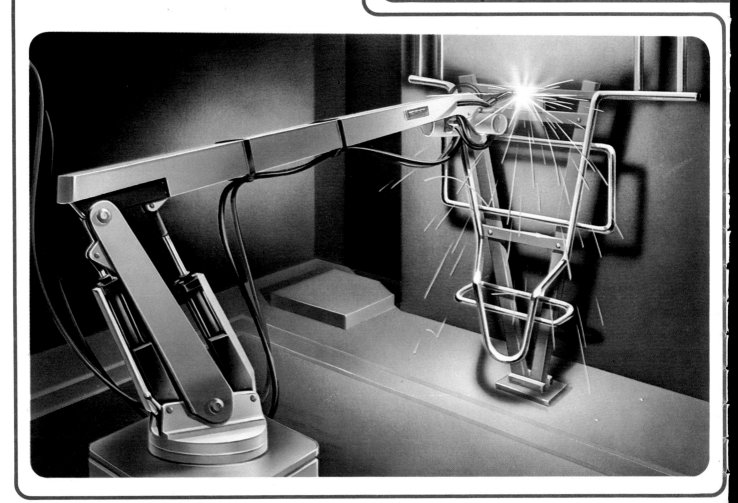

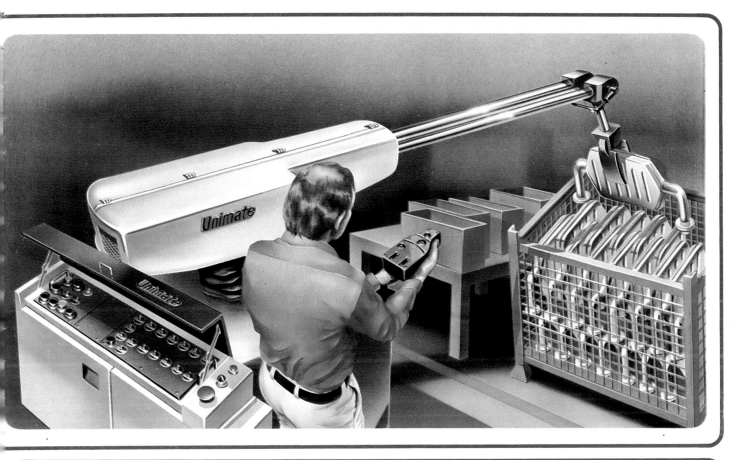

22 Can a robot learn to paint?

A robot that has been taught by a teach control unit moves in a series of jerks. However, in order to teach a robot to spray paint over an object like a chair, it must be able to move more smoothly. So a robot learns to paint by a method called continuous path teaching.

The robot's computer memory is set to 'record'. The operator then uses the robot arm to spray a sample object. In this way, the robot arm is led through all the right movements and they are recorded in its computer memory.

The operator has to work out the best way to paint the object, which may be difficult if it is on a moving conveyor belt. He may need to do it several times before he is satisfied. Then, once the right set of movements is finally recorded, the robot arm will paint all similar objects in exactly the same way.

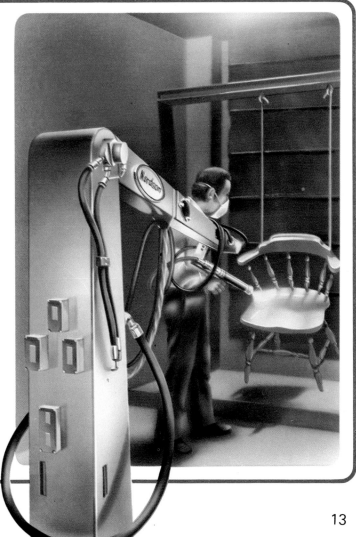

23 What is remote control?

A remote control machine has to be operated by a human at all times, even though the operator can be some distance away. Such a machine cannot learn and is not a true robot. However, most remote control machines have robot parts. They may have powerful arms, or grippers and, using television cameras and other sensors, some can feed back information to their operators.

The submersible *Consub*, shown here, is used for exploring the seabed and inspecting underwater constructions, such as oil rigs.

24 Why is remote control useful?

Remote control machines are used in places and situations where it is dangerous for humans. They can go deep under the sea, into very hot or cold areas and far out into space. They can be operated electrically, through cables, or, if some distance away, they may be controlled by radio signals.

The machine shown here is called Mobot. It handles dangerous radioactive chemicals while its operator stays safely in another room. It carries television cameras and microphones and its grippers can handle delicate glass containers. It can even be programmed, like a robot, to do a particular job over and over again.

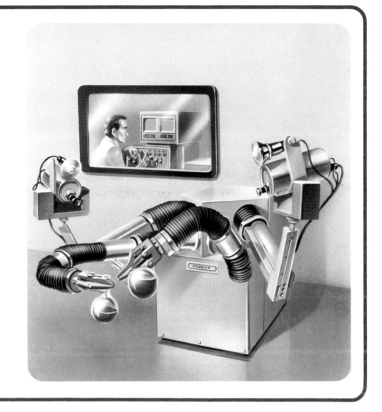

25 Can we feel by remote control?

This Man Mate manipulator is designed for working with very hot pieces of metal. The operator sits in an air-conditioned cab, while the temperature outside the cab may be unbearably hot. One of the special things about this machine is that it has feedback. Sensors in its gripper feed information to the controls, so the operator feels how much power the gripper is using.

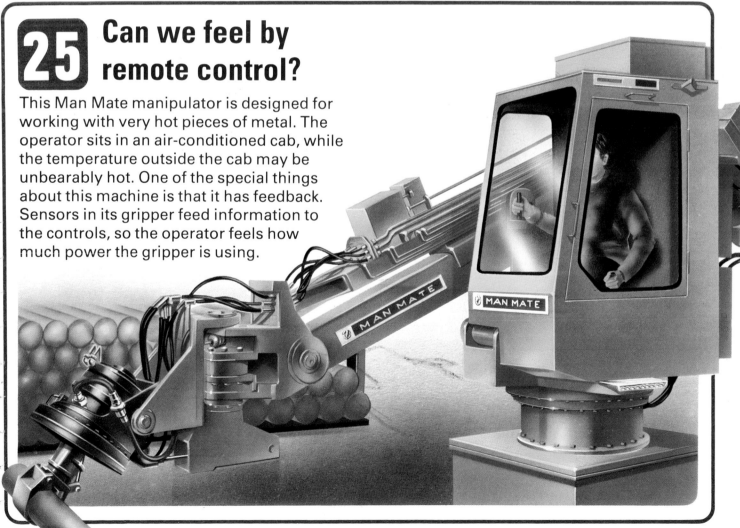

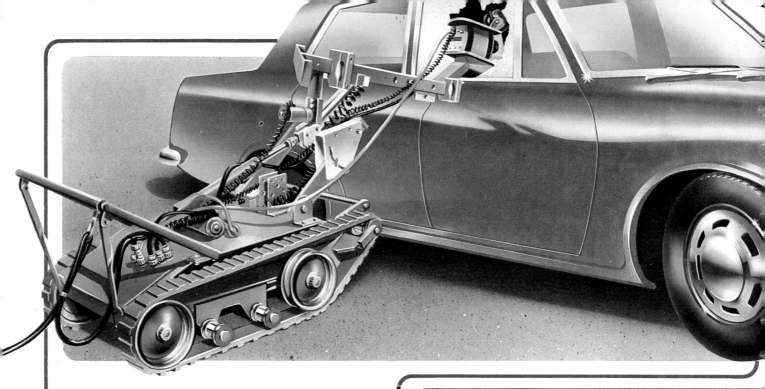

26 What does Wheelbarrow do?

Wheelbarrow belongs to the British Army. It is used for finding and dealing with booby-trap bombs in cars. Wheelbarrow is operated through an electric cable and has a television camera and several other tools. From a safe distance the operator can guide it to a suspect car, make it break open a window and put its camera inside. If there is a bomb in there, Wheelbarrow's operator can move the bomb around by remote control and may be able to defuse it. If not, the bomb can be exploded from a distance.

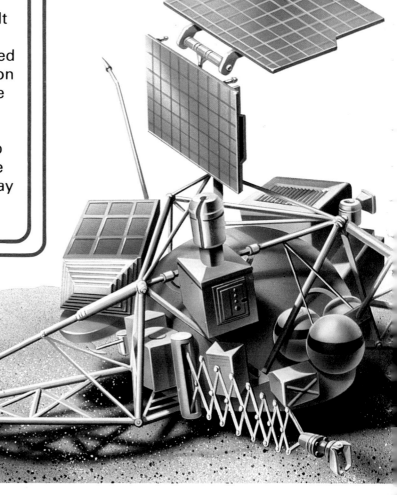

27 Did robots explore the Moon?

The first explorers on the Moon were all unmanned spacecraft, such as the Russian *Luna* probes and the American *Surveyor* craft. Unmanned spacecraft are, again, not true robots, as they are controlled by radio signals from Earth. However, they do have on-board computers and these give them a certain amount of self-control. The 1967 *Surveyor 3*, shown here, dug trenches in the lunar soil and photographed the results.

In 1970, *Luna 17* landed the first of the two *Lunokhod* rovers, which travelled across the surface, sending back information.

28 How was Mars explored?

The journey to Mars takes 11 months, which is too long, at present, for a manned expedition. Remote control spacecraft, however, can make the journey. The first probe, the Russian *Mars 1*, passed close to Mars in 1963. It was followed by the American *Mariner* probes and two more *Mars* probes.

The first real landings on Mars were achieved by the two *Viking* craft (one is shown here) in 1976. They both carried scoops and biological testing equipment and it was hoped that they might find some signs of life on Mars. But they found no evidence that there was ever any life on the planet.

29 What does Sim One do?

Sim One is not a robot, but he is an almost life-like computer-controlled model of a human. He is used to help teach student doctors. His heart beats, he breathes and his eyes blink. His skin looks and feels like human skin and he can be made to appear to have a number of different illnesses. He then responds to drugs and other forms of treatment. Machines like Sim One may help us to make real androids one day.

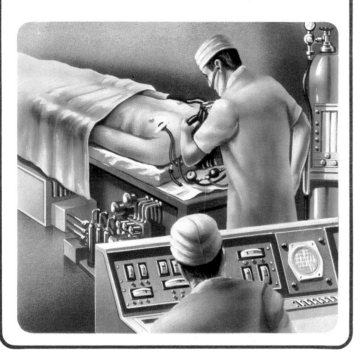

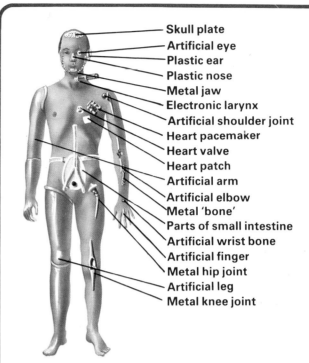

- Skull plate
- Artificial eye
- Plastic ear
- Plastic nose
- Metal jaw
- Electronic larynx
- Artificial shoulder joint
- Heart pacemaker
- Heart valve
- Heart patch
- Artificial arm
- Artificial elbow
- Metal 'bone'
- Parts of small intestine
- Artificial wrist bone
- Artificial finger
- Metal hip joint
- Artificial leg
- Metal knee joint

30 Is Bionic Man possible?

The Bionic Man of television uses very advanced electronic devices in place of his damaged limbs and eyes. Such things are not yet possible in the real world, but doctors are gradually learning how to replace more and more parts of the human body with artificial ones like those in the picture above. Perhaps one day there will be real bionic people.

31 How do artificial arms work?

A modern artificial arm is made up of a series of linkages powered by small electric motors, gas pressure or liquid pressure. Some arms are controlled by signals from the wearers' own muscles. Others are controlled by small computers. An arm developed recently in Sweden seems to give its wearers some sort of touch feedback, although the hand has no touch sensors.

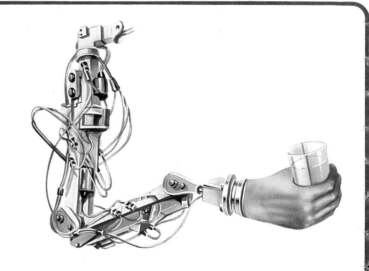

32 What is an exoskeleton?

An exoskeleton is a machine that can be worn by humans to increase the power of their movements. The Hardiman exoskeleton shown here is strapped on to the wearer. Using the exoskeleton a person can pick up an object weighing over 400 kilogrammes. Unfortunately, the machine cannot tell the difference between deliberate and unintentional movements, and can get out of control.

33 What is the Walking Lorry?

The Walking Lorry was built in America in 1969. It stands over three metres high and moves on its four legs with amazing gracefulness. It is controlled by an operator who crouches inside its cab. The four legs copy the movements of the operator's own arms and legs.

Like the Hardiman exoskeleton, the Walking Lorry greatly increases its operator's movements. It is equipped with sophisticated feedback devices so that the operator feels exactly what the machine is doing. With practice the operator can feel that the machine's legs are his own and that he is walking along on his hands and feet with immense strength – with a flick of his wrist he can toss aside a wooden beam as if it was a matchstick.

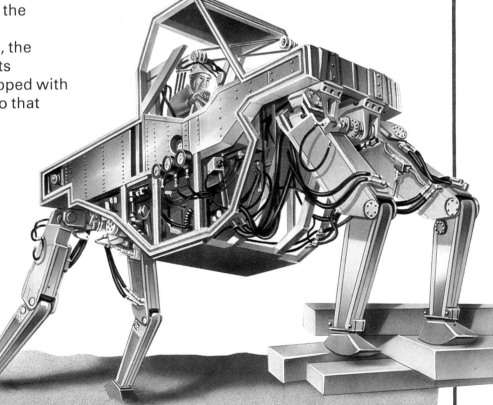

34 What are second generation robots?

The new, or second generation of industrial robots are smaller and more advanced than the lumbering first generation types (see page 10). Among the first of these robots is Unimation's PUMA, above, which can put together parts of small machines with an accuracy of one tenth of a millimetre.

As engineering improves, it may be possible to make the new robots more clever. Future robots should have cameras so they can see, and touch sensors so they can feel. Better computer languages will have to be developed so that supervisors can 'talk' to robots through keyboards, and the robots can reply through printers. It may also be possible for robots to be able to find faults in their own systems and to point out these faults to their supervisors. Finally, factory robots of the future should be able to move about and be able to perform more than one task without having to be reprogrammed.

35 How can a robot see?

It is easy enough to link a television camera to a robot's brain. But it is more difficult to make the brain recognize what the camera sees. The best way of doing this is to store a picture of an object in the robot's memory. When the robot is operating it can then compare what its camera sees with the picture in its memory. When both pictures are the same the robot will recognize the object and be able to react accordingly.

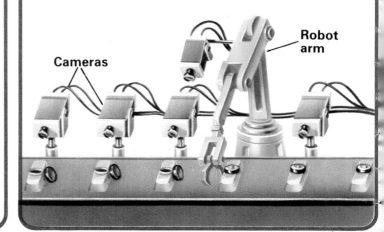

How can robots feel?

Robots can be made to feel by giving them some sort of touch sensors. The simplest type is just a switch that comes on when the robot is touching something. Some industrial robots now have pressure gauges that tell their computer if they are holding an object too tightly or have put it in the right place. The robot hand shown here has 384 sensors and can feel the shape, size and texture of the glass it is holding.

 ## Can you talk to a robot?

Robots can be given 'ears' (microphones) and 'voice boxes' (loudspeakers). There are even some machines that can make sounds like human speech. But getting a robot to understand speech is more difficult. At present, a few machines can only recognize a small number of words. We will probably need keyboards and printers to be able to 'talk' to robots for some time to come.

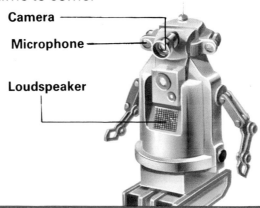

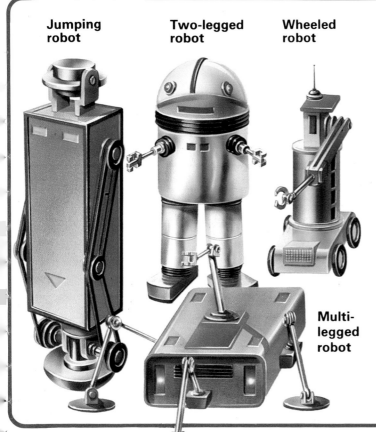

Jumping robot **Two-legged robot** **Wheeled robot** **Multi-legged robot**

 ## Can robots walk about?

A robot that moves must also be able to keep its balance. Wheels and tracks are more steady, or stable, than legs, but wheels need very smooth ground and even tracks cannot travel over every obstacle. Scientists have experimented with machines that have anything from one to eight legs. The more legs a robot has the more stable it is, but more legs means using more energy. At present, four or six legs seem to work best.

Tracked robot

39 Can a robot brain think?

No program exists at the moment that can make a computer think like a human. But programs do exist that allow a computer to make simple decisions. Some of the most advanced are computer chess games.

A good human chessplayer makes each move after having decided what possible moves may follow it. No one knows exactly how chessplayers do this, but they probably base their decisions on about 5000 rules. A simple example is: If 'A' happens and 'B' happens then 'C' will probably happen next.

Similar rules are now being used in some computer programs. A human expert teaches a computer the rules needed to help solve particular problems, like helping doctors diagnose illnesses, for example. In the future, perhaps, it may be possible to teach robot brains the very complicated rules that we use to understand the things we see and hear around us.

40 What can WABOT do?

WABOT-1 is an experimental Japanese robot. It was built to find out how robots might move about, and how they could be made to react to events and objects around them. WABOT-1 looks like a human, in that it has two legs, and two arms with hands. It has balance sensors and can walk across a room without falling over. It also has eight touch sensors in each hand, television cameras, a microphone and a loudspeaker.

When given a command, its 'voice' can say that it has understood. It has enough control to be able to pour water from one glass into another. And, when instructed to, it can 'see' an object, walk to it, pick it up and carry it to the right place.

41 How clever was Shakey?

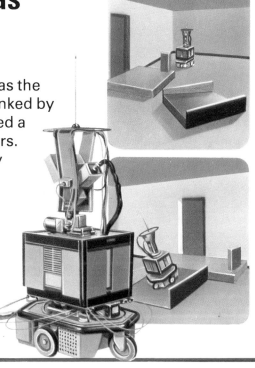
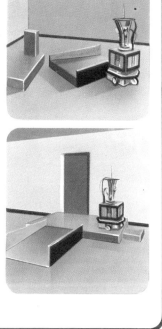

Shakey was first built in 1968, and was the first complete robot system. It was linked by radio to its computer brain and carried a television camera and various sensors. Working in its special rooms, Shakey could move wooden ramps about to allow it to reach shaped boxes and could then pile the boxes up in groups as instructed.

Shakey was very clever in its special surroundings. However, to be able to move around in the outside world, a robot will need to be much more clever than Shakey.

42 Are robot mice intelligent?

Electronic 'mice' are built to help scientists find out more about how to teach a machine to learn for itself. To help it find its way through a maze, a 'mouse' is given different sensors so it can recognize walls and gaps.

Some mice are programmed to obey certain rules, such as 'always keep a wall on the right'. Others are left to find their way by trial and error. Interestingly, 'trial and error' mice are sometimes faster than programmed mice. Electronic mice are not intelligent – their behaviour is more like instinct. However, by using such machines, scientists may one day be able to make robots behave like living things.

43 Could robots run factories?

More and more robots are being used in factories. There are over 10,000 industrial robots in Japan and over 5000 in the United States. New robots are being designed all the time. For example, in Japan, engineers are experimenting with a robot that can study an engineer's drawing and then assemble actual parts to match the drawing.

There are already factories in which whole stages of some manufacturing processes are handled by robots. So it is quite reasonable to suppose that, in time, robots could do all the jobs in a factory. A central robot brain would organize all the different types of jobs, and the robots to do them. These would include pick and place robots, assembly and transport robots and even cleaning and repair robots. Only a few people would be needed as supervisors in such a factory.

The most useful type of robot factory would be one that handled dangerous materials. However, in the near future, we will probably not want to have too many robot factories, as there may not be enough jobs left for people.

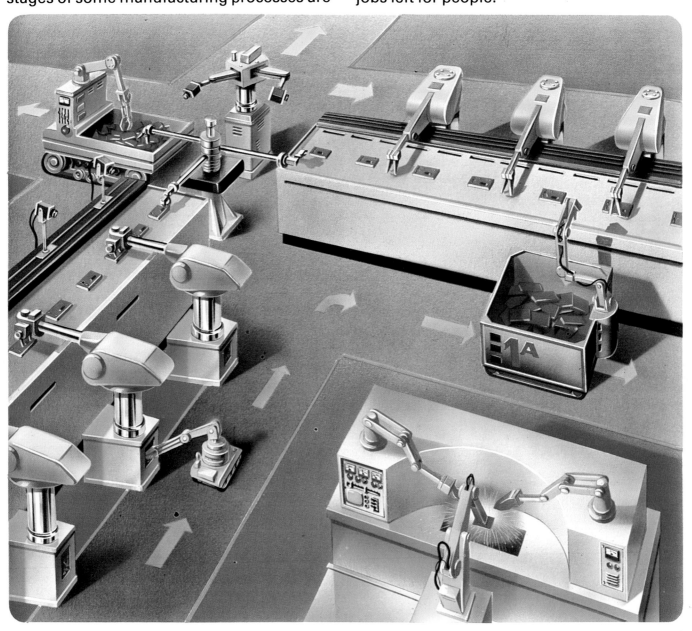

44 Could robots work on farms?

Japanese scientists are now trying to build robots that can be used in farming and forestry work. However, farms are not as easily organized as a factory process. It is much more difficult to teach a robot to decide whether an apple is ripe or a cow is healthy than it is to teach it to pick up a particular piece of machinery.

Even so, robots may be used on farms of the future, particularly for work that is already being done largely by machines — robot combine harvesters, milking machines and robot ploughs, for example.

Another use for robots might be in underwater fish farming. If the fish were kept in air bubble cages, robots could pass easily through the 'walls', and could be programmed to feed the fish at regular intervals. Other robots could harvest the fish and take them to an underwater factory.

45 How will robot cars work?

Modern cars are noisy and dangerous. Thousands of people are killed on the roads every year, so the idea of safe, driverless, robot cars is an exciting one.

The first robot cars will probably be used in cities. Travellers will park their own cars on the edge of a city and walk to the nearest autotaxi stop. Once there, a traveller will press a 'call' button and the first empty autotaxi in the stream of traffic that is passing will pull in. Inside the autotaxi, the passenger will press buttons to tell the autotaxi where to go and it will travel safely along its route, following an invisible track to its passenger's destination.

Eventually, robot cars may be possible on all roads. They will follow cables in the ground to stop them going off the road. Each car will have microwave radar, lasers and other equipment to help its robot brain to 'see' other cars on the road, and to adjust its speed to wet or icy surfaces. It will find its way to a particular place by picking up coded radio signals from 'bleepers' by the side of the road.

46 Will we have robots at home?

Robots work best in well-organized places where nothing ever changes. Few people have, or even want their homes as well-ordered as they would need to be for robots. How, for example, would you teach a cleaning robot to tell the difference between an old newspaper and the book you left on a chair? Also, it would be a great nuisance to have to reprogram your robot every time you moved a piece of furniture. And just imagine how dull food prepared by a cooking robot might be.

So, although the idea of having robots to do such work as cleaning and washing up seems like a good one, they would not really be very useful. We would have to organize our lives to suit them!

47 Could we have robot pets?

We might well find human-like robots in our homes difficult to live with, but robot pets would be much easier and could even be useful. A robot pet might be able to carry on a simple conversation and could play games, such as chess or ludo. Its memory could be used to remind its owner of things like shopping lists, people's birthdays and telephone numbers.

K-9 (*Dr Who*)

Muffitt (*Battlestar Galactica*)

48. Will we use robot weapons?

Robot weapons already exist. A Cruise missile flies at tree-top level, finding its way by comparing the ground below with the map in its computer program. There is also a robot torpedo, called Sting Ray, which has been programmed to recognize and 'home in' on the sound of an enemy ship's engine.

In the future, there may be robot fighters and bombers in the air and robot tanks on the ground. But they will probably also be controlled in some way by human operators. There may even be armed satellites in space. Using high-powered lasers, they could 'knock out' other satellites and blow up intercontinental missiles in mid-flight.

49. Could robots repair spacecraft?

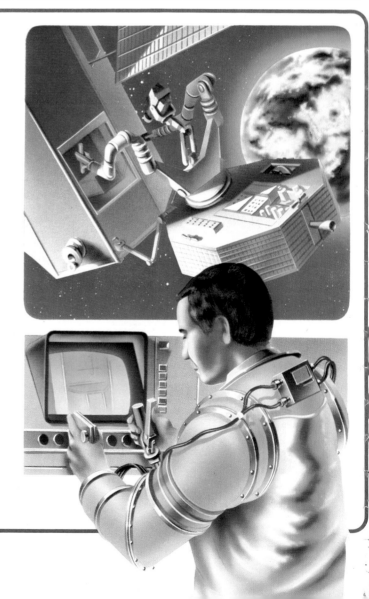

Repairing spacecraft in space may soon become very important. But the computer programs we have at the moment would only allow robots to do very simple jobs. However, it is possible that people could do the work by remote control.

The JPL/Ames Arm was originally made for use on spacesuits, but it can be used just as well as a robot arm. A pair of these arms on a remote control space vehicle could carry out the actual repair work. The robot arms would be controlled by an operator based on a nearby spaceship. The operator's own arms would be inside another pair of JPL/Ames Arms. The operator would watch the repair vehicle on a television screen. Every movement made by the operator's arms would be copied by the robot arms. Touch sensors on the robot arms would let the operator feel exactly what the repair vehicle was doing. The operator would feel as if he was doing the work himself.

50 Could robots build space cities?

Robots will probably do almost all of the building work in space. Each robot will be carefully programmed to do one or more particular jobs and all the work will be controlled by a 'master' robot brain on a spaceship.

The first space robot of this sort has already been designed. It will be used to put together the large cross-beams of space stations. The beams will be carried up from Earth by space shuttles.

If whole cities are ever built in space, robots will be used for mining the building materials from other planets, as well as for actually putting the cities together.

Robot Maze Games

Put a sheet of tracing paper over the maze board so that you can mark your robot's moves with a pencil. Start with Game 1 — each game is a bit more difficult.

The aim is to get the robot into the centre square, the arrow or 'V' shows the direction the robot must use to enter that square. The robot can move up or down, left or right, but cannot move in a diagonal direction. It can cross back over squares it has been in before, but must not land on the same square twice in the same game.

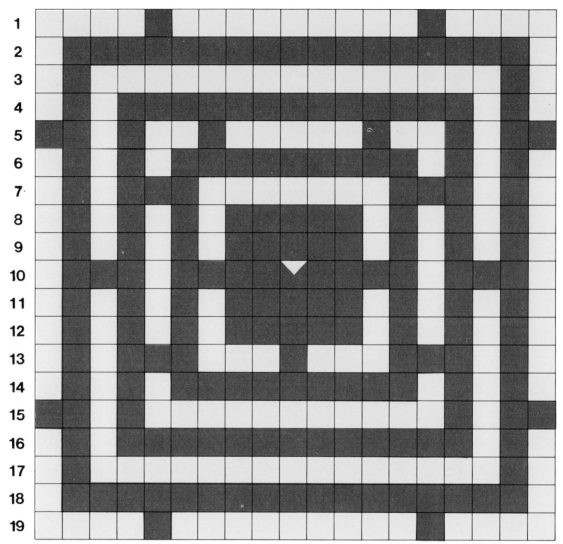

GAME 1
Program a robot to get on to the centre square in just 10 moves. The robot must make alternate moves of 3 squares and then 2 squares, or 2 squares and then 3. You must start outside the playing area and make the robot enter at one of the 8 red squares around the edges. The robot may only land on the red squares, but can move through the yellow squares.

GAME 2
Starting on square A1 or S1 program your robot to get on to the centre square in 8 moves. The robot must make alternate moves of 3 squares and then 2, or 2 squares and then 3. It may only land on the yellow squares, but can move through the red squares.

GAME 3
Starting on square A1 or S1 program your robot to get onto the centre square in 8 moves. The robot must make alternate moves of 1 square and then 4, or 4 squares and then 1. It may only land on the yellow squares, but can move through the red squares.

GAME 4
Starting on square A19 or S19 program your robot to get onto the centre square in 11 moves. The robot must make alternate moves of 3 squares and then 2, or 2 squares and then 3. It may only land on the yellow squares, but can move through the red squares.

GAME 5
Starting on square A19 or S19 program your robot to get onto the centre square in 9 moves. The robot must make alternate moves of 4 squares and then 1, or 1 square and then 4. It may only land on the yellow squares, but can move through the red squares.

The Multiple Robot Game

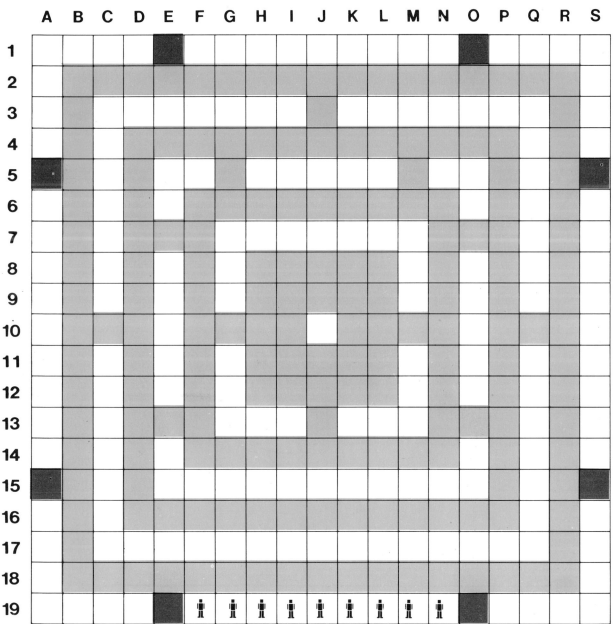

There are 9 robots on squares F19, G19, H19, I19, J19, K19, L19, M19 and N19. The aim is to get each robot onto its corresponding square at the top of the board, so F19 must get to F1, G19 to G1, etc. Each robot must make alternate moves of 3 squares and then 2 squares, or 2 squares and then 3.

No robot can land on a square that any other robot has previously landed on, except at the starting line (Row 19) and the finishing line (Row 1) – so program your robots carefully. The robots can only land on the red or white squares, but they can pass through the blue squares. They can move up or down, or left or right, but cannot move diagonally. They can also double back through squares they have already passed, but must not land again on the same square.

Use tracing paper and a pencil to mark off your robots' moves so you know which squares have been landed on. You may need several goes at this game.

INDEX

A
Androids 6, 18
Arithmetic unit 8
Artificial arms 10, 18
Artificial parts 18
Asimov, Isaac 7
Automata 3, 4

B
Balance sensors 22
Binary system 9
Bionic Man, The 6, 18
Bionics 6, 18

C
Čapek, Karel 3
Central processing unit (CPU) 8
Circuit boards 8
Computer chess 22
Computers 4, 5, 6, 8, 9, 11, 12, 17, 18, 22, 23, 28
Computer terminal 8
Consub 14
Continuous path teaching 13
Cruise missiles 28
C-3PO 3
Cybermen 6
Cyborgs 6

D
Daleks 7
Degrees of freedom 10
Digital computer 8, 9

E
Electronic mice 23
Exoskeletons 19

F
Farming robots 25

Feedback 4, 5, 12, 15, 18, 19
First generation robots 10, 20
Flowchart 9

H
Hardiman exoskeleton 19

I
Industrial robots 3, 5, 10, 12, 20, 21, 24
Input unit 8

J
JPL/Ames Arm 28

K
Keyboards 8, 20, 21
K-9 27

L
Laws of Robotics 7
Luna probes 17
Lunokhod rovers 17

M
Man Mate manipulator 15
Mariner probes 17
Mars probes 17
Mechanical toys, *see* Automata
Memory, Computer 8, 12, 13, 20
Mobot 15
Muffitt 27

O
ON/OFF 6

P
Performing robots 6

Pick and place robots 11, 12
Position sensors 12
Program 5, 9, 12, 22, 28
Programming robots 9, 11
PUMA 20

R
Remote control machines 3, 14, 15, 16, 17, 28
Robot cars 26
Robot pets 27
Robot weapons 28

S
Scribe, The 3, 4
Second generation robots 20
Shakey 23
Silicon chip 8
Sim One 18
Sting Ray 28
Surveyor probes 17

T
Teach control unit 12, 13
Tobor the Great 7
Touch sensors 18, 20, 21, 22, 28

U
Unmanned spacecraft 17

V
Viking probes 17

W
WABOT-1 22
Walking Lorry 19
Wheelbarrow 16

C-3PO (page 3) by courtesy of Lucasfilm Ltd.